Mainstreaming the Environment

THE WORLD BANK GROUP
AND THE ENVIRONMENT
SINCE THE RIO EARTH SUMMIT

FISCAL 1995

Summary

THE WORLD BANK
WASHINGTON, D.C.

1818 H Street, N.W.
Washington, D.C. 20433, U.S.A.

Manufactured in the United States of America
First printing September 1995

The text is printed on recycled paper that exceeds the requirements of the 1988 guidelines of the U.S. Environmental Protection Agency, section 6002 of the Resource Conservation Recovery Act. The paper stock contains at least 50 percent recovered waste paper material as calculated by fiber content, of which at least 10 percent of the total fiber is postconsumer waste, and 20 to 50 percent of the fiber has been deinked.

Printed on Recycled Paper

ISBN 0-8213-3481-6
ISSN 1014-8132

Acknowledgments

This report was prepared by the Environment Department of the World Bank in close collaboration with regional environment divisions and country operations departments. The principal author was Jocelyn Mason. John Redwood III and John Dixon contributed to chapters 1 and 3, respectively; Lars Vidaeus and Stephen Lintner wrote chapter 2; chapter 5 drew on a background paper written by William Magrath of the Agriculture and Natural Resource Department; and chapter 6 drew on three background papers written by Alison Cave and Jerry Lebo (urban and transport, respectively) of the Transport, Water and Urban Development Department, and Rachel English (energy) of the Industry and Energy Department. Letitia Oliveira of the International Finance Corporation (IFC) wrote chapter 7. The Environment Department team included Kerstin Canby, John Kellenberg, Marian Mabel, and Mary McNeil. Clare Fleming provided advice for the illustrations and cover design. Andrew Steer provided guidance.

Foreword

It is now three years since almost all countries of the world endorsed the concept of environmentally sustainable development at the Earth Summit in Rio de Janeiro, Brazil. This report documents how the World Bank Group has sought to be an active partner in implementing the "Rio imperatives." It is divided into three parts. Part 1 explores progress in activities specifically targeted toward improving the environment. This includes an analysis of the Bank's growing loan portfolio of environmental projects—now $10 billion for 137 projects in 62 countries—and of the Bank's role as an implementing agency of the Global Environment Facility (GEF) and of the Montreal Protocol. Part 2 asks a broader question: How are environmental concerns being incorporated throughout *all* of the Bank's activities? It provides a preliminary "green accounting" of the $67 billion that the Bank has committed in the past three years. Part 3 documents the environmental programs of the International Finance Corporation (IFC) and the Multilateral Investment Guarantee Agency (MIGA).

While documenting the good progress that has been made to date, the report also tries to point to the areas where extra effort is now needed. In a sense, the Bank is now entering its "third generation" of environmental reforms. The first, in the 1987–92 period, was characterized by a major focus on reducing potential harm from Bank-financed projects and, specifically, the codification of environmental assessment (EA) procedures. The second might be termed the "post-Rio boom." It was characterized by a great expansion in the Bank's environmental capacity, and an aggressive effort to respond to the exploding demand for Bank assistance in environmental management.

The third generation is now under way. It is characterized by three main thrusts. The first is an overriding emphasis on on-the-ground implementation. The second is a major effort to move "upstream" from project-specific concerns in order to incorporate the environment into sectoral and national strategies. The third, undergirding all our activities, is a stronger focus on people and on social structures to find solutions and make development more sustainable.

Andrew Steer
Director
Environment Department
The World Bank

Overview

JAN POST

Ecuador is endowed with one of the highest concentrations of biological diversity per unit area in South America, of which the tree frog pictured above is only one example. To protect the country's biological heritage, the government of Ecuador has established a national system of protected areas with fifteen conservation units of global importance. Fourteen areas are located on the continental territory of Ecuador, and one, Galápagos National Park, is in the Galápagos Islands about 1,000 kilometers off the coast. A five-year GEF-funded project will support the management plans of eight of these sites, as well as outreach activities focusing on conflict resolution among key target groups and public awareness campaigns.

A quiet revolution has been under way during the first half of the 1990s, as environmental sustainability has become a theme of policymaking around the world. Environmental problems remain acute in most countries, and in many parts of the world continue to escalate. But there has been a quantum increase in awareness of the magnitude of the problem and of the need for action in the past few years, with the 1992 United Nations Conference on Environment and Development (UNCED) in Rio de Janeiro, known as the Rio Earth Summit, capturing the growing consensus and dramatically accelerating the momentum for change. More than 100 nations are trying to build environmental concerns into their planning processes, and in about half those nations substantial changes in policy and investment priorities are evident. Member countries have sought the World Bank Group's support in this endeavor—through financing, advisory services, and donor coordination.[1] This report attempts to document this growing partnership.

As the World Bank Group has assisted its member countries in making development sustainable, so has it undergone its own "greening." A series of operational policies addressing the environmental aspects of Bank activities has been put into place. Operational departments are strengthening their capacity for environmental analysis; the number of technical environmental staff has increased fivefold since the end of the 1980s; and a new Vice Presidency of Environmentally Sustainable Development has been created.

This report assesses this progress, giving special note to the acceleration of activities in the three years since the Rio Earth Summit, and paying particular attention to the most recent year, fiscal 1995. While capturing the substantial gains that have been made, it also shows that the journey is unfinished (see box 1). In the next three years, the World Bank Group will need to focus its attention in particular on the implementation of the investments and policies that have been designed over the recent past.

The report is divided into three parts. Part 1 addresses progress in those areas of financing and technical assistance that are targeted *specifically* toward the environment. This includes that portion of the Bank's active portfolio, currently $10 billion, directed primarily at helping borrowing countries improve their environmental management (chapter 1).[2] It also includes Bank-supported programs to address global environmental problems, often financed by the Global Environment Facility (GEF) and the Montreal Protocol (chapter 2). Finally, it covers the recent analytical, research, and policy work undertaken by the Bank on environmental issues (chapter 3).

However, while the specific environmental programs are important, of equal importance is the extent to which environmental concerns are

Box 1. The Unfinished Agenda

The report describes substantial progress in the way in which the World Bank assists its client countries in addressing environmental concerns, but it also notes that the agenda is unfinished. The following are among the challenges expected for the coming three to five years:

- *Implementing the portfolio of environmental projects.* The majority of projects in the currently active portfolio of $10 billion in sixty-two countries are new, and $8 billion is scheduled to be disbursed over the next few years. Ensuring that these projects continue to perform well and that lessons from their innovative approaches are learned and disseminated will be major challenges.
- *Moving beyond project-specific environmental assessment.* The past few years have seen a remarkable deepening in the effectiveness of the Bank's environmental assessment work. It is necessary now to broaden the focus to apply these techniques at the sectoral and regional levels. Such approaches are already being piloted to good effect.
- *Improving the monitoring of on-the-ground impacts.* Information on environmental trends remains very weak. This undermines the ability of policymakers to set priorities and monitor the impact of programs. The Bank's new efforts to develop and implement indicators of progress will need to be deepened.
- *Addressing the social dimensions of environmental management.* Ensuring that stakeholders are involved in the design and implementation of projects, that cultural issues are taken into account, and that the social costs and benefits of interventions are carefully discussed and addressed will need to receive greater attention in the future.
- *Developing and applying global overlays.* There is a growing urgency for integrating global environmental objectives into national strategies and action plans for sustainable development (that is, global overlays). Extensions will be made to country economic and sector work with developing country partners.

incorporated into the *entire* portfolio of the Bank's activities. This is the focus of Part 2. It includes, for example, the extent to which potentially adverse environmental and social impacts of Bank-financed projects are assessed and mitigated (chapter 4) and the way in which the concept of environmental sustainability is reshaping the Bank's portfolio in major sectors of activity such as agriculture and forestry (chapter 5) and infrastructure (chapter 6).

Part 3 of the report examines how the International Finance Corporation (IFC) and the Multilateral Investment Guarantee Agency (MIGA) address environmental concerns in their private sector activities.

Targeted Programs for the Environment

The Bank's active portfolio of loans whose primary objective is to strengthen environmental management now stands at $10 billion for 137 projects in 62 countries (table 1). Of this amount, $5.6 billion has been committed in the three years since UNCED, and $1.1 billion for 21 projects was added in fiscal 1995 (see figure 1). New borrowers for the environment in fiscal 1995 included Honduras, Latvia, Lebanon, Lithuania, the Russian Federation, Thailand, Trinidad and Tobago, and the Organization of Eastern Caribbean States (see figure 2 for a map representing all countries with Bank-supported environment projects).

The Portfolio of Environmental Projects

Pollution management and urban environment projects account for about 61 percent of total lending for the environment, and 40 percent of the number of environmental projects in the active portfolio (table 1). Such projects typically include four types of components: capacity building for environmental management, including training, staffing, monitoring, and policy studies; resources for on-lending to enterprises and agencies for improved environmental management; direct investment

Table 1. Active Portfolio of Environmental Projects as of June 30, 1995

Focus of the project	*Number of projects*	*Number of countries*	*World Bank contribution (loan or credit; billions of dollars)*	*Total project cost (billions of dollars)*	*Average size of loan or credit (millions of dollars)*
Pollution management (brown agenda)	55	31	6.1	16.1	110.0
Natural resources (green agenda)	62	39	3.2	5.8	52.0
National institution building	20	19	0.7	1.2	33.0
Total	137	62[a]	10.0	23.1	73.0[b]

a. Total number of borrowing countries; some countries have more than one project.
b. Average size of loan for the whole portfolio.
Source: World Bank data.

Figure 1. The Active World Bank Environmental Portfolio

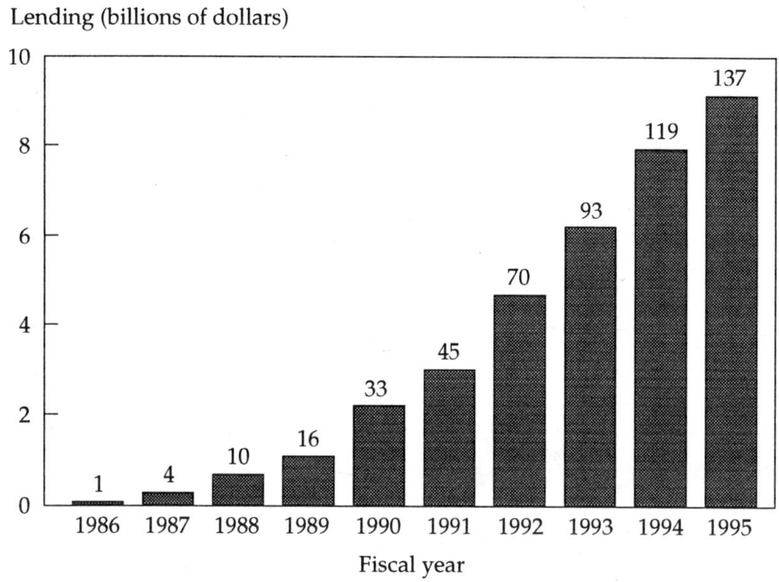

Note: Numbers at top of columns represent total number of active environmental projects, and include projects until the fiscal year in which they were closed or canceled.

in pollution prevention and treatment facilities; and support for research on new technologies and improved policy design. In fiscal 1995, fourteen new loans for environmental management and pollution control were made, with objectives varying from introducing clean fuel in Thailand to preventing industrial pollution in India and Poland; improved municipal water treatment and coastal zone management in the three Baltic countries; improved solid waste management and sanitation in Burkina Faso, Lebanon, and Sri Lanka; and emergency treatment of an oil spill in the Russian Federation.

Rural environmental projects, now under way in thirty-nine countries, cover a range of issues, including managing forests and conserving biodiversity; establishing or strengthening national parks and other protected areas; introducing integrated pest management; and investing in soil conservation, watershed rehabilitation, and integrated river basin management. New projects in fiscal 1995 in Venezuela (park protection), Pakistan (afforestation), India (forest management), and Brazil (Rain Forest Pilot Program) illustrate the new approaches in addressing rural environmental problems. Prepared in partnership with affected groups

Figure 2. Countries with Environmental Projects

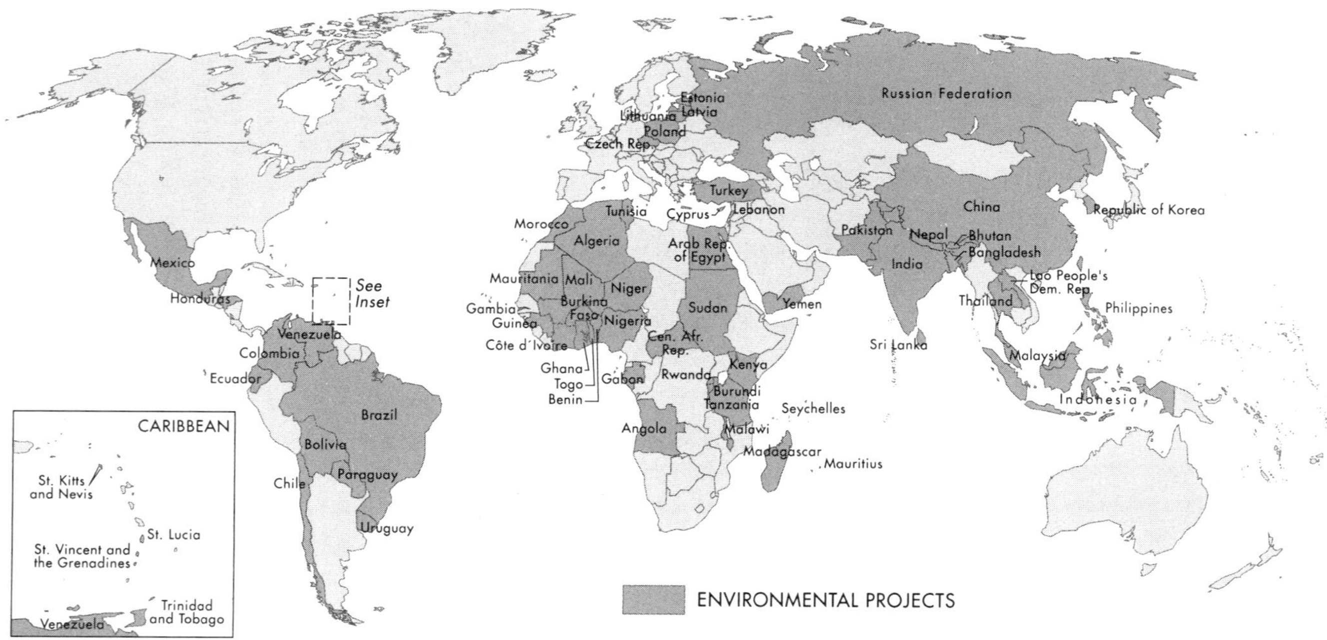

and nongovernmental organizations (NGOs), emphasizing the need for alternative livelihoods and changed incentive structures, and recognizing administrative capacity constraints, these projects will seek to ensure better livelihoods for local residents as a result of nondamaging activities.

Environmental institutions projects, directed toward strengthening national and local environmental management capacity (covering both pollution and rural issues), are under way in twenty countries. Two-thirds of these projects have been approved in the past three years, and in fiscal 1995 new borrowers included Benin, Honduras, Trinidad and Tobago, and the Russian Federation. The project in the Russian Federation, supported by a $110 million loan, is of particular note because it will help the Russian government to establish an Environmental Framework Program throughout the nation. Measures to strengthen public involvement in environmental decisionmaking have also increased in Bank-supported environmental institution-building projects.

Regional and Global Programs

The World Bank's work as coordinator or financier of regional or global activities that address transboundary environmental problems has grown markedly over the past few years.

COOPERATION AT THE REGIONAL LEVEL. Current examples at the regional level include the following:

- The Mediterranean Technical Assistance Program (METAP), which over the past four years has supported nearly 100 capacity-building and investment-preparation activities in the region. Its partners are currently preparing its third program cycle.
- The Baltic Seas Environment Program, for which the Bank has developed seven investment projects for financing by the end of calendar year 1995.
- The Danube River Basin Program, which this year completed, with assistance from the Bank, a strategic action plan for improving water quality and addressing hot spot problems. This plan will guide Bank and other coordinating agency activities in the Danube Basin.
- The Aral Sea Program, for which a priority action plan, prepared in collaboration with the GEF implementing agencies—the United Nations Environment Programme (UNEP), the United Nations Development Programme (UNDP), and the World Bank—has been agreed by the governments of Kazakhstan, the Kyrgyz Republic, Tajikistan, Turkmenistan, and Uzbekistan. Implementation is expected to begin in early 1996.

- The Lake Victoria Program, for which the governments of Kenya, Tanzania, and Uganda agreed in 1994 to prepare a joint investment program with support from the Bank and the GEF.

In addition to these activities, the past years have seen active engagement by the Bank in programs for the Caspian Sea, the Red Sea, and the International Coral Reef Initiative. In March 1995 the Bank issued, in collaboration with the World Conservation Union (IUCN) and the Great Barrier Reef Marine Park Authority, a major study identifying global priorities for marine protected areas (World Bank, IUCN, and the Great Barrier Reef Marine Park Authority 1995).

ADDRESSING GLOBAL PROBLEMS. The past year has seen the first meetings of the parties to the Convention on Biological Diversity and the Framework Convention for Climate Change, and rapid progress toward ratification of the Desertification Convention.[3] The Bank is actively supporting the implementation of each of these conventions and over the past year has presented strategy statements as to how, through its policy, advisory, and financing roles, and as implementing agency to the Global Environment Facility (GEF) and Montreal Protocol, it can make an active contribution.

The past year has seen the completion of the GEF pilot phase and the start of GEF 1, a program of $2 billion addressing global warming, biodiversity conservation, ozone depletion, and international waters. In the past twelve months, a new GEF project cycle has been introduced, and a comprehensive operational strategy setting priorities in the four focal areas has been drafted. In parallel, draft guidelines on social assessment in GEF and Montreal Protocol projects have been prepared to guide project designers.

As of July 1995 the GEF Council had endorsed World Bank Group financing (using GEF resources) of sixty-three projects, totaling $557 million, of which fifty-two projects ($454 million) were from the pilot phase, and eleven projects ($103 million) from GEF 1. The number of Bank-supported Montreal Protocol projects (for addressing ozone depletion) has risen this year to twenty-four in twenty different countries with a total funding of $145 million.

In helping to address global environmental problems, the Bank seeks to apply two principles: integration of these global concerns into national development policymaking and increased effectiveness of scarce grant resources through financial leverage and market-based approaches. The past year has witnessed some important innovations in Bank-supported projects, including introduction of sectoral approaches toward ozone-depleting investments emphasizing unit-cost reductions in China; experimentation with market-based approaches to resource allocation for phaseout of ozone-depleting substances in Chile; the development of rapid appraisal techniques for setting biodiversity priorities; partnership with the government of Norway on demonstration projects for market-based carbon emissions offsets in Poland and Mexico; market research on the feasibility of renewable energy and biodiversity venture capital funds; and initiation of a research program of "global overlays," assessing the opportunities, costs, and benefits of integrating global concerns into domestic policymaking.

The Intellectual Journey

The Bank's rapidly growing financing program has been supported by an active program of analytical and policy work. This work falls into three categories: preparation of country and regional strategies; cross-country analysis of best practice and policy; and the development and piloting of new methodologies.

COUNTRY AND REGIONAL STRATEGIES. The Bank has advocated and supported the preparation of National Environmental Action Plans (NEAPs) since 1992. These are prepared by national governments, although the Bank often provides analytical support and helps to mobilize resources. As of July 1995, sixty-three countries had completed NEAPs (forty-seven IDA borrowers and sixteen IBRD borrowers). NEAPs are also nearing completion for many of the new members of the Bank, such as the countries of the former Soviet Union, as well as for older but recently inactive borrowers, such as Algeria, Haiti, Lebanon, and Viet Nam. In March 1995 the Bank issued a review of experience to date with national environmental planning (World Bank 1995j). It concluded that these exercises were beginning to have a genuine and positive impact on policy, although the quality of the analysis and recommendations is variable. In identifying best practice to date, the review concluded that the most effective strategies explicitly address three questions:

- How to set priorities
- How to design cost effective interventions
- How to build capacity and remove obstacles to implementation.

The best strategies are those that effectively draw upon and weave together rigorous analysis and participatory consensus-building. Many exercises to date have emphasized one at the expense of the other—undermining either the quality of the recommendations or the likelihood of effective implementation.

The Bank has also sought to assist in defining strategies at the regional level. The Environmental Action Programme for Eastern and Central Europe, coordinated by the Bank and approved by governments in 1993, is now being implemented. This was followed in 1994 and 1995 by preparation of regional reports on East Asia, Sub-Saharan Africa, and North Africa and the Middle East.

DISTILLING BEST PRACTICE AND DEFINING BANK POLICIES. Accompanying these country and regional strategies has been a growing array of cross-country studies on special topics of concern to the Bank's clients. The Bank, like all others in the development community, remains on the steep part of the learning curve. While in many areas of policymaking the theory is clear, institutional constraints often call for adjustments to textbook solutions. Best-practice studies have been undertaken for most aspects of the environmental agenda—including pollution manage-

Box 2. Ten Hallmarks of the New Environmentalism

As developing countries struggle to forge development paths that provide both prosperity for their citizens *and* good stewardship of the environment, they are rejecting the old-fashioned development-versus-the-environment paradigm in favor of a new environmentalism that recognizes economic development and environmental sustainability as partners. Some of the principles underlying this new approach are listed below. These also underlie much of the Bank's environmental support to its member countries.

- *Set priorities carefully.* Not all problems can be addressed at once. Priorities need to be set using a combination of modern valuation techniques and participatory input from citizens and experts.
- *Go for win-win options first.* Although tradeoffs exist, there is often scope for improving the environment by adopting policies and investments that are justified for reasons other than the environment. These free lunches, however, need to be exploited carefully.
- *Promote cost-effectiveness.* As demonstrated by recent work in countries like Chile and Mexico and in regions such as Eastern Europe, huge opportunities exist for saving resources by carefully analyzing costs. This has rarely been the practice in industrial-country policymaking.
- *Use market incentives where feasible.* A growing body of evidence, some supported by Bank projects, points to the gains from market-based instruments. Chile, Hungary, Malaysia, Poland, and Thailand are examples.
- *Economize on administrative and regulatory capacity.* Administrative and enforcement capacity is often as scarce as money. In these cases, blunter instruments (with fewer points of intervention) and self-enforcing policies (for example, deposit-refund schemes) can help.

ment, sanitation, water management, natural habitat protection, forest policy, and so on. Chapter 3 of this report describes the lessons emerging from such studies and how they are being incorporated into the Bank's training, advisory, and project work. In certain key areas such findings are codified into Bank policy statements. In the recent past this has occurred for forest policy and water management. In addition, in fiscal 1995 an "Industrial Pollution Prevention and Abatement Handbook" (World Bank 1995g) was issued, and new insights into policies that work are disseminated widely through working papers and dissemination notes.

Although it is not possible to summarize the findings of all recent studies, some common themes are emerging. A new environmentalism is spreading throughout the Bank's member countries, supported in part

So can involving local communities and NGOs in enforcement and monitoring.

- *Set realistic standards and enforce them.* The old approach of adopting Western standards, failing to meet them, and thus undermining the credibility of the government's commitment is giving way to a new approach whereby realistic and phased targets are set and enforced.
- *Work with the private sector.* Much of the investment required to improve the environment will come from private businesses and farmers. A shift in attitude is required (and in many places is occurring), with an emphasis on incentives, consultation, good business citizenship, and public disclosure of good as well as bad corporate performance.
- *Recognize that public involvement is crucial.* Success or failure often depends more on the commitment of stakeholders, including local communities, than that of the central government. Protecting natural habitats, forests, and even river basins requires genuine community involvement in planning and implementation. Many successful examples are emerging.
- *Build constituencies for change.* Sustained improvements in environmental policymaking, which will usually involve losers as well as winners, are more likely with a well-informed public and active support from citizens. The most effective governments are building such constituencies for change—through education and awareness programs and through promotion of an active civil society.
- *Incorporate environmental concerns from the outset.* Prevention is usually much cheaper than cure. Environmental concerns need to be built not only into the beginning of each investment, but also into the fabric of economic policymaking, including tax and regulatory policy, and even into the national accounts.

by the active portfolio of Bank-supported projects. Some of the emerging hallmarks of this new environmentalism are presented in box 2. These principles, while seeming rather obvious today, certainly were not so even five to ten years ago, which in itself indicates how rapidly the new approach is spreading.

DEVELOPING NEW METHODOLOGIES AND INDICATORS. The analytical tool kit available to policymakers in many areas of the environment remains very weak, and the Bank is devoting substantial energy to developing new practical methodologies for identifying problems and designing efficient solutions. Examples during the past year are the refinement and piloting of new approaches for identifying and projecting pollution emissions; participatory decisionmaking for priority setting at the local level; the analysis of the impacts of road building on natural habitats; the identification of the impacts of economywide policy reforms on the environment; the enhancement of water resource management, and the assessment of the social impact of Bank-supported investments.

Finally, a major effort has been made over the past two years to improve the quality of environmental data and indicators. A major publication, *Monitoring Environmental Progress,* has just been published (World Bank 1995i). Given the poor quality of most environmental data, it proposes action to improve indicators in the coming years and computes a number of new indicators of progress. For example, it presents the first systematic adjustments of net national savings rates and national wealth calculations, taking into account natural as well as manmade capital. Follow-up work will be pursued jointly by the Bank's Environment and Development Economics departments.

Greening the Entire Portfolio

Part 2 of this report broadens the coverage to ask how the Bank's *entire* lending portfolio is being adjusted in the light of new knowledge and concern about the environment. In the three years since the Rio Earth Summit (fiscal years 1993–95), the Bank has committed $67 billion for investments around the world. How are the lessons of the new environmentalism being factored into these investments?

A simple—and rough—breakdown of Bank lending since Rio is presented in table 2. Part 2 of the report first looks at how environmental assessment procedures are increasingly affecting the design of those investments that could potentially harm the environment. It then explores two broad sectors of activity—agriculture and forestry, and infra-

Table 2. World Bank Lending since Rio—A Simple Accounting

Type of lending	*Billions of dollars*	*Percentage*
Total lending	67	100
Environment projects[a]	6	9
"Win-win" projects[b]	20	30
Investments with potentially significant harmful impacts on the environment[c]	13	19
All other lending	28	42

a. See annex C for a listing of environmental projects. Adjustments have been made elsewhere to avoid double counting. Dollar amounts have been rounded.

b. "Win-win" projects are those that promote both economic growth and improvements to the environment but are not classified in the table as specifically environmental. "Win-win" projects that are also categorized as A are not included.

c. Based on projects categorized as A by regional environment divisions.

structure—which, combined, account for 60 percent of the Bank's lending in the past three years.

Making a Difference with Environmental Assessments

All Bank projects must be screened for their potential environmental impacts. Those expected to have adverse impacts that may be sensitive and irreversible are considered to be in category A and are subject to a full environmental assessment. Category B projects are those whose impacts are expected to be less significant; they are subject to some environmental analysis. Category C projects are not expected to have any negative environmental impacts. Among the investment projects approved over the past year, 23 (10 percent) were classified as category A; 81 (36 percent) were classified as category B; and the remaining 124 (54 percent) were classified as category C.

In June 1995 the Bank completed its "Second Review of Environmental Assessment," covering the roughly 600 projects that have been screened in the three years since UNCED (World Bank forthcoming b). The review concluded that environmental assessments (EAs) are now firmly rooted as part of the Bank's normal business activity, and are continuing to improve in quality as Bank staff and borrowers learn from experience. EAs are now effecting changes in project design as a matter of course and are getting better at systematically considering investment alternatives, at including monitoring and management plans for mitigation measures, and at improving public consultation.

A number of areas for future work emerged from the review. One example is the use of sectoral EAs (SEAs), which move the EA upstream in the design process and enable EAs to affect project design more effectively early on.

Agriculture, Forestry, and Natural Resources

A growing recognition of the environmental challenges of both extensification and intensification and of the nonmarket values of forests, wetlands, and other natural habitats has had a major impact on lending in these sectors over recent years. Chapter 5 explores this evolution in terms of three broad issues:

IMPROVED LAND MANAGEMENT. In giving greater emphasis to ensuring that increases in yield are *sustained*, the Bank's support is shifting in three important ways:

- *Move to least-cost, quick-return methods of soil and moisture conservation* by supporting existing farmer practices, as in the Loess Plateau in China for example, and by developing easily implemented practices, such as water-harvesting techniques using rock bunds in Africa.
- *Encourage market-based land reform* by attacking the nexus of policies that favor large-scale farms at the expense of small family farms—policies that by restricting the poor's access to land lead to poverty, inefficient resource allocation, and natural resource degradation. Land reform, such as that under way with Bank support in Colombia, will become an important theme of Bank activity.
- *Shift toward decentralized rural development* to benefit the rural poor, take advantage of local knowledge, earn the support of populations affected by development, and benefit from the flexibility to adapt to local circumstances. Thus the Bank is encouraging community-based resource management, as in a number of projects for livestock management in Africa. In addition, providing finance to poor farmers, particularly women, has become an important way both of helping them break the cycle of poverty and of enabling them to invest in sustainable natural resource management.

IMPROVED WATER MANAGEMENT. In implementing the 1993 Bank policy for managing water resources (World Bank 1993c), Bank support is continuing to shift toward four imperatives:

- *Support a comprehensive view of water management issues.* The Bank is encouraging policies that are sustainable from economic, social, and environmental perspectives. This is particularly the case in fostering cooperation among riparian countries.

- *Emphasize improvement of the efficiency of existing irrigation systems before creating new ones.* Since issuing the new policy, the Bank has focused its irrigation portfolio on rehabilitating existing systems and their management. This has meant increasing user participation in the design of irrigation and drainage projects, as well as institutional strengthening for water management.
- *Encourage decentralization and local management of water resources.* In an effort to improve efficiency and accountability of public sector irrigation management agencies, more than half of Bank irrigation projects approved over the last two years included user participation in project management.
- *Impose charges for using water that reflect its value as a scarce resource.* More than one-third of irrigation water in developing countries is typically lost between farm gate and crops, leading to water scarcity, soil salinity, and groundwater pollution. A growing number of Bank projects support the implementation of equitable user charges for water.

IMPROVED FOREST MANAGEMENT. Since the adoption of its new forest policy in 1991 (World Bank 1991a), the Bank has significantly increased its lending to the sector and given special emphasis to four new principles. A progress report in fiscal 1995 (World Bank 1994i) highlighted action and continued room for progress in each of the principles above.

- *Rectify market and policy failures that encourage deforestation and inhibit sustainable land use,* by promoting sector policy and institutional reform and by seeking to bring a more realistic assessment of resource values into decisionmaking.
- *Give special emphasis to expanding public participation in forest resource planning and management and mobilizing private sector resources and skills,* by encouraging greater stakeholder participation in forest management and planning, and by increasing support for alternative livelihoods for the rural poor living within forest boundaries.
- *Expand and intensify the management of areas suitable for sustainable forest production, including establishing plantations to reduce pressures on the existing base of forest resources where the scope to do so is sound from a social, environmental, and economic perspective,* by low-cost technical improvements as well as farmer training.
- *Expand the designation of forest areas as parks and reserves by adopting and encouraging borrower governments to pursue a precautionary approach to forest use, particularly in tropical moist forests,* by increasing the share of lending for this purpose.

Making Infrastructure Environmentally Sustainable

Investments in energy, transport, and urban infrastructure pose the greatest threat to the environment, but also offer the greatest scope for improvement on past practice. Chapter 6 documents important shifts under way in each of these sectors.

ENERGY. Building on the principles laid out in 1993 in major policy papers on energy efficiency and electric power (World Bank 1993a, 1993e), the Bank continues to strengthen its support for a two-pronged strategy to make the energy sector more environmentally sound.

- *Improve demand and supply-side efficiency.* In the fiscal 1993–95 period, twenty loans have supported power sector reforms to improve efficiency in production and transmission and in demand and end use. Innovative approaches in a diverse range of countries—including Bulgaria, Colombia, Honduras, Pakistan, and Papua New Guinea—are being monitored closely. In addition, free-standing components to encourage energy efficiency have also increased substantially over the last three years. These include support for both the more traditional supply-side measures as well as innovative approaches to peak period use and other measures. The Bank's support for energy efficiency is having a demonstrable impact: projects to develop coal-fired electricity plants supported by the Bank, for example, are shown to be significantly more efficient, in terms of the quantity of carbon emitted for each unit of energy produced, than the developing-country average.
- *Encourage clean technologies.* However successful developing countries are in realizing efficiency gains in energy production, transmission, and use, these alone will not be enough to meet future demand in an environmentally acceptable way. The Bank is therefore increasingly supporting clean sources and technologies: shifting to natural gas and clean coal for power generation, improving the quality of automotive fuels, investing in emission control technologies, and supporting the development of renewable sources of energy. The growing portfolio of renewable energy projects includes three geothermal projects in the Philippines; a solar photovoltaic and wind-farm project in India; a biomass energy project in Mauritius; and (under final preparation) new renewable energy projects in India, Indonesia, and Sri Lanka. In 1995 the Bank inaugurated a Solar Initiative, whose objective is to accelerate the pace at which commercial and near-commercial

renewable energy applications reach the marketplace. The GEF is an important partner with the Bank in some of these new initiatives.

TRANSPORT. Since UNCED, Bank activities to balance transport activities and environmental concerns have focused on two critical issues.

- *Address transport-related pollution.* In many heavily polluted developing-country cities, transport already accounts for 95 percent of lead and carbon emission and 60 to 70 percent of nitrogen and hydrocarbons. In response, the Bank is sharply expanding its support for investments designed to reduce pollution—both through managing the demand for private road transport and through direct support for controlling vehicle emissions. The Mexico Transport Air Quality Management Project, which was designed on the basis of careful ranking of the cost-effectiveness of alternative measures and which began implementation this year, captures this new approach. The Bank has also actively advocated and supported the total phaseout of lead from gasoline. It is currently assisting countries in both the technical aspects of refinery conversion and the incentive and public education aspects of the transition.
- *Reduce urban congestion.* While the Bank has traditionally addressed this problem by investing in traffic management and support for transport planning, it is now increasingly promoting the use of demand management instruments such as pricing signals for congestion, fuel, and public transport. Parking charges in Tunis, Tunisia, and rationing of off-street parking in Pusan, Republic of Korea, are examples. The Bank is also exploring a wider range of options including, in some megacities, capital-intensive investments such as rapid rail systems, which until recently have not been economically viable. In São Paulo, Brazil, for example, the Integrated Transport Project will support the construction and operation of a rail connection between the suburban train systems and the subway.

A draft of a new paper (World Bank, TWU 1995) on sustainable transport was completed this year. The main thrust of this paper is to shift away from mitigating existing environmental problems toward preventing them in the first place. This paper will be followed by a program of training and research dissemination for Bank staff and borrower countries.

TOWARD SUSTAINABLE CITIES. The three years since UNCED have seen a major effort to implement the 1991 policy paper on urban development

(World Bank 1991b). The policy underscored the importance of sound urban environmental management both as a prerequisite for urban productivity and as essential to the health and well-being of the urban poor. Recent Bank work in urban areas has been guided by a number of principles, which include the following:

- *Promote integrated resource-based project design.* A new generation of projects, especially those dealing with water resources, recognize the need to coordinate interventions around a resource, rather than around a jurisdiction or sector. For instance, the Estonia Haapsalu and Matsalu Bays Environment Project includes both water supply and treatment and management plans to improve beaches for tourism and protect endangered wetland habitat. Similarly, the Organization of Eastern Caribbean States (OECS), partially supported by the GEF, will improve urban solid and liquid waste management with the objective of protecting sensitive coastal zones.
- *Encourage decentralization to municipalities and communities.* As urban centers grow and become more complicated to manage, it is increasingly clear that municipal governments are more likely than national governments to find effective solutions to urban management. Moreover, throughout Bank operations, and in the urban sector in particular, stakeholder participation is proving to be an essential component of successful project implementation. The Asunción Sewerage Project in Paraguay, for example, decentralizes activities to improve water and sewerage services. By doing so, Asunción will also enable and encourage private sector participation.
- *Incorporate cost recovery and demand management.* Subsidies for urban services not only are costly but rarely benefit the poor whose plight is used to justify their existence. Bank-financed projects continue to put an emphasis on applying appropriate tariffs and cost-recovery mechanisms for urban services. Recent examples include water supply projects in Brazil, China, Latvia, and Lebanon. In addition, in urban environmental management, as in environmentally sustainable development more generally, the participation of the private sector is of critical importance to mobilize private financial resources as well as its innovative capacities. Bank-supported urban infrastructure projects for privatization, or for the intermediary stage of corporatization, have been approved over the last three years throughout the developing world, including examples from the Baltic countries to Peru.

- *Build partnerships for the urban environment.* Finally, the Bank has increasingly reached out to join partners in addressing the issues of urban pollution. Notable among these efforts are the Metropolitan Environmental Improvement Program with the UNDP, the Urban Management Program with the United Nations Center for Human Settlements (UNCHS) (Habitat) and the UNDP, as well as such regional programs as the Mediterranean Technical Assistance Program with the Commission of the European Communities, the European Investment Bank, and the UNDP.

Strengthening the Private Sector Role: The IFC and MIGA

Developing countries have increasingly recognized the critical role the private sector must play in financing sustainable development. A rapid growth in IFC and MIGA activity since the Rio Earth Summit has reflected this. Concurrently, moreover, the nature of the Bank Group's support for the private sector has changed to reflect the fact that, in industrial and developing countries alike, businesses have become more aware of the environmental effects of their activities. As the IFC's emphasis on environmental issues has grown, the way in which it does business has changed in two key areas: it has become more proactive in using its strengths to encourage the private sector to address environmental issues, and its internal capability to address environmental matters has been strengthened.

PROACTIVE PROGRAMS. Taking an increasingly proactive approach to supporting the private sector in environmentally sustainable development, the IFC has focused on the development of primarily environmental projects. It has also improved guidance on environmental matters to small and medium-size firms, advised financial institutions on environmental risk management, and explored innovative financing mechanisms to address environmental issues, such as targeted investment funds.

Although expanding the portfolio of environmental projects has proved slower than anticipated, two projects approved last fiscal year, one in Argentina and one in Mexico—both to improve and expand water and wastewater services—have met with considerable success and are having an important demonstration effect. Two environmental projects were approved this year, including a second investment in the Argentina project.

In the area of the global environment, the IFC, in cooperation with GEF and the Multilateral Fund for the Montreal Protocol, is preparing five projects, two of which have already been approved by IFC management. Three of these, in Egypt, the Slovak Republic, and Turkey, will help reduce chlorofluorocarbon emissions; a fourth will support the manufacture of compact fluorescent light bulbs in Poland; and the fifth will provide investments through financial intermediaries to small and medium-size firms for work on climate change or biodiversity.

In addition, the IFC assists small and medium-size companies with training and capacity building to introduce them to environmental management systems and techniques for avoiding and minimizing waste and pollution, as well as to IFC's environmental requirements.

ORGANIZATION AND PROCEDURES. Since the Rio Earth Summit, the IFC has made significant changes in the way it organizes its operations to address environmental matters. An Environment Division, which has grown from two members to eighteen in the last three years, supports the IFC's investment departments as well as MIGA. The director of the Technical and Environment Department now reports directly to the IFC's executive vice president. Furthermore, to achieve a better integration of environmental activities within IFC operations, a Vice-President of Operations has assumed corporate oversight for environmental matters. Like the Bank, the IFC screens its operations for their environmental impact, and those that are deemed to have a potentially significant impact on the environment are subject to a full environmental assessment. Since the IFC adopted new requirements on disclosure of environmental information in 1994, information on these and other projects is now available worldwide through the Bank's Public Information Center (PIC). In addition to disclosing environmental information, the IFC also requires sponsors of its category A projects to consult with affected local groups and interested parties during the preparation of the EA.

Notes

1. The World Bank Group, as used in this report, refers to the International Bank for Reconstruction and Development (IBRD) and its affiliates: the International Development Association (IDA), the International Finance Corporation (IFC), and the Multilateral Investment Guarantee Agency (MIGA). IBRD and IDA together are commonly referred to as the World Bank, or simply the Bank.

2. All dollar amounts are current U.S. dollars, except where noted. A billion is 1,000 million.

3. The Desertification Convention is the United Nations Convention to Combat Desertification in Those Countries Experiencing Serious Drought and/or Desertification Particularly in Africa.

Annex A. Projects with Primarily Environmental Objectives under Implementation Approved since Rio

This annex lists projects with primarily environmental objectives and currently under implementation that were approved in the three years since the Rio Earth Summit. They are ordered according to the environmental area they address: urban environmental management and pollution control (the brown agenda), natural resource management (the green agenda), or environmental institution building.

Annex A has been created for this summary. Tables A.1–A.3 correspond to tables 1.1–1.3 in chapter 1 of the full-length report.

Table A.1. Projects for Urban Environmental Management and Pollution Control, Fiscal 1993–95
(millions of dollars)

Fiscal year and country	*Project name*	*Loan/credit (L/C)*	*World Bank financing*	*Total project cost*
1993				
Brazil	Water Quality and Pollution Control Project—São Paulo/ Paraná	L	245	494
	Minas Gerais Water Quality and Pollution Control Project	L	145	308
China	South Jiangsu Environment Protection Project	L	250	584
Egypt	Private Sector Tourism Project	L	130	784
India	Renewable Resources Development Project	L + C	190	440
Korea, Rep. of	Kuangju and Seoul sewerage Project	L	110	530
Mexico	Transport Air Quality Management Project	L	220	1,087
Turkey	Bursa Water Supply and Sanitation Project	L	130	258
Total			1,420	4,485
1994				
Algeria	Water Supply and Sewerage Rehabilitation Project	L	110	170
Brazil	Espirito Santo Water Supply and Coastal Pollution Management Project	L	154	308
China	Shanghai Environment Project	L	160	457
Ecuador	Mining Development and Environmental Control—Technical Assistance Project	L	14	24
Estonia	District Heating Rehabilitation Project	L	38	65
Indonesia	Surabaya Urban Development Project	L	175	618
Mexico	Northern Border Environment Project	L	368	762
	Second Solid Waste Management Project	L	200	416
	Water and Sanitation II Project	L	350	770
Togo	Lomé Urban Development Project	L	26	29
Total			1,595	3,619

Fiscal year and country	*Project name*	*Loan/credit (L/C)*	*World Bank financing*	*Total project cost*
1995				
Burkina Faso	Urban Environment Project	C	37	50
China	Liaoning Environment Project	L	110	351
Estonia	Haapsulu and Matsalu Bays Environment Project	L	2	8
India	Industrial Pollution Prevention Project	L + C	168	330
Korea, Rep. of	Ports Development and Environmental Improvement Project	L	100	1,107
	Waste Disposal Project	L	75	305
Latvia	Liepaja Environment Project	L	4	21
Lebanon	Solid Waste/Environmental Management Project	L	55	135
Lithuania	Klaipeda Environment Project	L	7	23
OECS countries	Solid Waste Management Project	L + C	12	51
Poland	Katowice Heat Supply and Conservation Project	L	45	93
Russian Federation	Emergency Oil Spill Recovery Project	L	99	140
Sri Lanka	Colombo Environmental Improvement Project	C	39	49
Thailand	Clean Fuels and Environmental Improvement Project	L	90	370
Total			843	3,033
Total since UNCED, fiscal 1993–95			3,858	11,137
Active projects approved before fiscal 1993			2,217	4,942
Total active portfolio			6,075	16,079

Table A.2. Projects for Natural Resources Management, Fiscal 1993–95
(millions of dollars)

Fiscal year and country	*Project name*	*Loan/credit (L/C)*	*World Bank financing*	*Total project cost*
1993				
Egypt	Matruh Resource Management Project	C	22	31
Gabon	Forestry and Environment Project	L	23	38
India	Uttar Pradesh Sodic Lands Reclamation Project	C	55	80
Indonesia	Integrated Pest Management Project	L	32	53
Pakistan	Northern Resource Management Project	C	29	40
	Fordwah E. Sadiquia Irrigation and Drainage Project	C	54	71
Seychelles	Environment and Transport Project	L	5	7
Tunisia	Second Forestry Development Project	L	69	148
Turkey	Eastern Anatolia Watershed Rehabilitation Project	L	77	121
Total			366	589
1994				
Bhutan	Third Forest Development Project	C	5	9
China	Forest Resource Development and Protection Project	C	200	356
	Loess Plateau Watershed Rehabilitation Project	C	150	259
Colombia	Natural Resource Managment Program	L	39	65
India	Andhra Pradesh Forestry Project	C	77	89
	Forestry Research Education and Extension Project	C	47	56
Indonesia	National Watershed Management and Conservation Project	L	57	488
Lao PDR	Forest Management and Conservation Project	C	9	20
Pakistan	Balochistan Natural Resources Management Project	C	15	18
Paraguay	Natural Resources Management Project	L	50	79
Poland	Forest Development Support Project	L	146	335

Fiscal year and country	*Project name*	*Loan/credit (L/C)*	*World Bank financing*	*Total project cost*
Tunisia	Northwest Mountainous Areas Development Project	L	28	50
Uruguay	Natural Resources Management and Irrigation Project	L	41	74
Total			864	1,898
1995				
India	Madhya Pradesh Forestry Project	C	58	67
Pakistan	Punjab Forest Sector Development Project	C	25	34
Venezuela	INPARQUES Project	L	55	96
Total			138	197
Total since UNCED, fiscal 1993–95			1,368	2,684
Active projects approved before fiscal 1993			1,825	3,079
Total active portfolio			3,193	5,763

Table A.3. Projects for Environmental Institutions, Fiscal 1993–95
(millions of dollars)

Fiscal year and country	*Project name*	*Loan/credit (L/C)*	*World Bank financing*	*Total project cost*
1993				
Bolivia	Environmental Technical Assistance Project	C	5	5
Chile	Environment Institutions Development Project	L	12	33
China	Environment Technical Assistance Project	C	50	70
Ghana	Environment Resource Management Project	C	18	36
Korea, Rep. of	Environmental Research and Education Project	L	60	97
Total			145	241
1994				
Gambia, The	Capacity Building for Environmental Management—Technical Assistance Project	C	3	5
Korea, Rep. of	Environmental Technology Development Project	L	90	156
Morocco	Environmental Management Project	L	6	11
Total			99	172
1995				
Benin	Environmental Management Project	C	8	9
Honduras	Environmental Development Project	C	11	13
Russian Federation	Environmental Management Project	L	110	195
Trinidad and Tobago	Environmental Management Project	L	6	11
Total			135	228
Total since UNCED, fiscal 1993–95			379	641
Active projects approved before fiscal 1993			289	536
Total active portfolio			668	1,177

Annex B. Projects with Primarily Environmental Objectives Approved in Fiscal 1995

This annex provides details on projects approved by the Bank's Board of Directors between July 1, 1994, and June 30, 1995, that have primarily environmental objectives. Table B.1 illustrates the broad range of environmental concerns addressed by Bank projects. The dollar figures provided for projects with environmental objectives represent the Bank share (IBRD loan or IDA credit), as well as the total project cost. Annex B corresponds to annex C in the full-length report.

Table B.1. Projects with Environmental Objectives Approved in Fiscal 1995

Country	*Project*	*Project description*
Africa		
Benin	Environmental Management ($8 million, IDA; $9 million, total project cost)	Allows the government to develop the national environmental management capacity required to implement and monitor effectively the priority actions identified in the National Environmental Action Plan (June 1993). Supports planning and implementation of policies and programs, monitoring and enforcement of legislation, strengthening of coordination mechanisms, development of information systems, and promotion of the integration of environmental concerns in socioeconomic development plans. Improves the awareness and understanding of environmental issues for various segments of the population and fosters the integration of environmental considerations in the education system.

(Table continues on the following page.)

Table B.1 *(continued)*

Country	*Project*	*Project description*
Burkina Faso	Urban Environment ($37 million, IDA; $50 million, total project cost)	Assists with the rehabilitation or improvement of priority primary drainage networks, the development of piped sewerage and on-site sanitation, the closing or upgrading of existing landfills and construction of new landfills, and the improvement of current toxic and industrial solid waste management systems. Aims to strengthen the central government's capacity for urban management.
Asia		
China	Liaoning Environment ($110 million, IBRD; $351 million, total project cost)	Assists the government in reducing pollution and improving operational efficiency through upgrading of technology, minimization of waste, and reuse. Strengthens local capabilities in monitoring and enforcement of environmental regulations, corporatization of the sewerage utilities, and establishment of a sound cost-recovery policy for pollution abatement services. Includes an environmental protection fund for eligible pollution control subprojects of industrial enterprises and enhances institutional development through technical assistance and training. Also includes a component to rehabilitate and conserve cultural heritage sites.
India	Industrial Pollution Prevention ($168 million, IBRD/IDA; $330 million, total project cost)	Promotes cost-effective abatement of industrial pollution through components that support institutional strengthening of state pollution control boards as well as investments by individual firms for pollution abatement with a focus on minimization of waste and cleaner production methods. Renews financing for common effluent treatment plants for treating wastewater and solid materials at industrial and other sites with a concentration of industries, particularly those of small size. A technical assistance component supports the development, diffusion, and transfer of technologies with environmental benefits.

Country	*Project*	*Project description*
India *(continued)*	Madhya Pradesh Forestry ($58 million, IDA; $67 million, total project cost)	Supports improvements for the Madhya Pradesh forestry sector, including innovative programs for participatory forest management. Involves forest regeneration and improved silvicultural practices; improved implementation, monitoring, and evaluation; a village resource development program based on participatory planning and the integration of forest management and protection with activities to generate alternative incomes to reduce pressures on the forest; forestry research, extension, and technology improvements; and biodiversity conservation through improved management of twelve protected areas while supporting the development of alternative resources or incomes for local communities.
Korea, Rep. of	Waste Disposal ($75 million, IBRD; $305 million, total project cost)	Assists the governments in Pusan and Chunbak Province in addressing environmental, institutional, and technological concerns regarding wastewater and specified waste disposal, the health hazards of surface and groundwater contamination, and the reuse of treated effluent. The Kusan waste disposal plant will ensure satisfactory management of most hazardous waste substances generated in industries.
	Ports Development and Environmental Improvement ($100 million, IBRD; $1,107 million, total project cost)	Provides comprehensive support for addressing environmental issues in port and harbor development. Includes facilities to collect, manage, and treat ship waste and to remove timber- and fish-processing plants that cause air, noise, and traffic pollution. Provides training and technical assistance to strengthen the environmental capability of the Korea Maritime and Port Authority and a program to monitor pollution, ecosystem productivity, biodiversity, and changes in the yields of living marine resources in the Yellow Sea.

(Table continues on the following page.)

Table B.1 *(continued)*

Country	*Project*	*Project description*
Pakistan	Punjab Forest Sector Development ($25 million, IDA; $34 million, total project cost)	Strengthens the Punjab Forestry Department's capabilities to assist local communities and the private sector in the development and conservation of forest and rangelands through natural, financial, and human resource management. Helps to disseminate technology and information required to promote farmers' involvement in nursery and timber production; develop and strengthen community organizations to manage scrub forests and rangelands jointly with public sector institutions; and implement a pilot component for the reorganization of timber plantations to enhance their environmental value.
Sri Lanka	Colombo Environmental Improvement ($39 million, IDA; $49 million, total project cost)	Focuses on improving solid waste services, controlling surface and groundwater contamination from wastewater discharge, and building up the capabilities of the local governments in municipal waste management. Promotes private sector participation in environment-related operations. Finances the construction of a sanitary landfill and compost facilities to replace open dumps and equipment for the collection, transport, and disposal of hospital wastes; wastewater collection systems in industrial zones; wastewater pollution abatement measures in the Beira Lake catchment area; and technical assistance.
Thailand	Clean Fuels and Environmental Improvement ($90 million, IBRD; $370 million, total project cost)	Supports the reduction of air pollutants attributable to petroleum fuels in Thailand by assisting the government in meeting reformulated gasoline and diesel oil specifications, improving its refinery operations through the installation of appropriate facilities and equipment to reduce refinery emissions and ensure safety.

Country	*Project*	*Project description*
Eastern Europe and Central Asia		
Estonia	Haapsalu and Matsalu Bays Environment ($2 million, IBRD; $8 million, total project cost)	Includes improvements in water and wastewater treatment in Haapsalu and promotes management of point and nonpoint source pollution in the catchment areas of Haapsalu and Matsalu Bays, with a view to reducing pollution in the Baltic Sea. Supports the planning and management of ecotourism and implementation of the management plan for the Matsalu State Nature Reserve. This project is part of the Bank's support for the Baltic Sea Environment Program.
Latvia	Liepaja Environment ($4 million, IBRD; $21 million, total project cost)	Includes water and wastewater investment components, which are complemented by an environmental management component. Restores and enhances water quality in the city of Liepaja and the northern portion of Lake Liepaja and promotes environmentally sustainable development, including tourist development, of the coastal zone and protected areas in and around Liepaja, Ventspils, and Talsi regions. This project is part of the Bank's support for the Baltic Sea Environment Program.
Lithuania	Klaipeda Environment ($7 million, IBRD; $23 million, total project cost)	Focuses on the city of Klaipeda. Seeks to improve water supply and sanitation services in the city as well as the operational efficiency and management of the Klaipeda State Water Supply Enterprise and to promote environmentally sustainable management and development of the Kursiu Lagoon and adjacent coastal areas. This project is part of the Bank's support for the Baltic Sea Environment Program.

(Table continues on the following page.)

Table B.1 ***(continued)***

Country	*Project*	*Project description*
Poland	Katowice Heat Supply and Conservation ($45 million, IBRD; $93 million, total project cost)	Addresses the environmental, health, and economic problems caused by soot and dust particulates generated by energy production. Enhances energy conservation and efficiency in the district heating sector by extending the life of existing district heating assets through rehabilitation and introduction of modern technologies and by reducing environmental pollution through investments in energy-efficient equipment and systems as well as by supporting a program to eliminate coal-fired, heat-only boilers.
Russian Federation	Environmental Management ($110 million, IBRD; $195 million, total project cost)	Provides support to the Russian Federation's Environmental Framework Program, which addresses environmental and natural resource management issues at federal, regional, and local levels and mainstreams them into the economic, social, and political adjustment process. Includes policy and regulatory support and environmental epidemiology; water quality management; hazardous waste management; the creation and capitalization of a National Pollution Abatement Facility for the funding of economically and financially viable pollution abatement projects; and the creation of a Center for Project Preparation and Implementation under the Ministry of Environmental Protection and Natural Resources.

Country	*Project*	*Project description*
Russian Federation *(continued)*	Emergency Oil Spill Recovery ($99 million, IBRD; $140 million, total project cost)	During 1994, one of the world's largest oil spills occurred in the Usinsk region of the Komi Republic as a result of leaks from the regional oil pipeline. The project assists the Russian Federation and the pipeline operator in stabilizing the oil in the spill area and preventing ecological damage in the Pechora River Basin; continues cleanup in an environmentally appropriate way; supports safe pipeline operations in the near term and evaluates the need for a replacement pipeline; and identifies and implements other measures to mitigate against possible oil spills in the future.
Latin America and the Caribbean		
Honduras	Environmental Development ($11 million, IDA; $13 million, total project cost)	Strengthens the governmental agencies' capacity in environmental and natural resource planning; policy, legal, and regulatory framework development; environmental monitoring; and the enforcement of environmental laws and regulations. Assists with the strengthening and implementation of a national system of environmental impact assessment that engages NGOs, communities, and the private sector in the review process. Develops the environmental management capacity of selected municipalities in a manner that will increase the involvement of affected organizations and individuals.
OECS countries	Solid Waste Management ($12 million, IDA and IBRD; $51 million, total project cost)	Improves existing national systems of solid waste storage, collection, and disposal; provides for investments in port reception facilities to address the problems of ship-generated solid waste and the rationalization of the existing institutional framework through the creation of solid waste management entities. At the regional level, provides technical assistance for sewerage master plans, sewerage and sewage treatment projects, and the preparation of an institutional framework for regional coordination.

(Table continues on the following page.)

Table B.1 ***(continued)***

Country	*Project*	*Project description*
Trinidad and Tobago	Environmental Management ($6 million, IBRD; $11 million, total project cost)	Establishes an environmental management agency that is responsible for the legal and regulatory framework covering all aspects of the environment in the country. Through technical assistance and training, helps to bring the newly created Environmental Management Authority to full operational capacity.
Venezuela	INPARQUES ($55 million, IBRD; $96 million, total project cost)	Strengthens the government's capability to manage Venezuela's national and urban parks, natural monuments, and wildlife refuges and reserves. In particular, the project strengthens conservation and protection of vulnerable areas; intensifies public environmental research, training, and education efforts, and improves the economic sustainability of the national parks and other protected areas.
Middle East and North Africa		
Lebanon	Solid Waste/ Environmental Management ($55 million, IBRD; $135 million, total project cost)	Completes the rehabilitation of the solid waste collection and disposal systems as envisaged under the National Emergency Reconstruction Program, improves the collection and disposal of hospital waste, strengthens the government's and the private sector's capabilities for solid waste management, and helps to develop a coastal zone management plan that will protect the Lebanese coast from further degradation.

Annex C. Projects with Major Environmental Components Approved in Fiscal 1995

This annex provides details on projects approved by the World Bank's Board of Directors between July 1, 1994, and June 30, 1995, that have major environmental components. Annex C corresponds to annex D in the full-length report.

Table C.1. Projects with Major Environmental Components Approved in Fiscal 1995

Country	*Project*	*Environmental component*
Africa		
Central African Republic	Livestock Development and Rangeland Management	Supports activities focused on rangeland management; grassroots capacity development, and the establishment of agropastoral areas or buffer zones that help to conserve the savanna ecosystems.
Côte d'Ivoire	Private Sector Energy	Reduces environmental impacts of energy production by supporting efficient gas-fired capacity and the use of indigenous natural gas.
	Municipal Support	Contributes to sustainable development through improvement and protection of the urban environment.
Ethiopia	National Fertilizer Sector	Supports the training of farmers and extension personnel in efficient and balanced use of nutrients so that losses of nutrients and other polluting elements into the environment are minimized; establishes arrangements to compile and maintain a data base on the impact of fertilizers and other materials on the environment; and establishes monitoring and testing activities at strategic locations for environmental protection.

(Table continues on the following page.)

Table C.1 ***(continued)***

Country	*Project*	*Environmental component*
Gabon	Transport Sector Technical Assistance	Provides training and seminars to sensitize actors in the transport and urban sectors to environmental considerations in the design, implementation, and monitoring of transport and urban projects, plans, and policies.
Ghana	Fisheries Subsector Capacity Building	Provides for the establishment of a monitoring, control, and surveillance system for marine and inland fisheries, as well as the long-term conservation, protection, and sustainability of the fishery resource.
	Mining Sector Development and Environment	Provides support to the Mining Commission to review existing monitoring and enforcement practices (including an assessment of jurisdictional responsibilities of sectoral agencies and the Environmental Protection Agency with regard to environmental monitoring); to develop guidelines and standards for implementing sectoral and environmental regulations; and to establish an environmental information system.
Madagascar	Second Irrigation Rehabilitation	Improves environmental management through the development of techniques for environmental assessment of irrigation rehabilitation and the reinforcement of local capacity for such assessment; improves the drainage of stagnant waters, which will have a beneficial effect on the accumulation of salinity and mineral deposits in the soils.
Malawi	National Water Development	Ensures the protection and management of water resources and of aquatic and riparian environments through institutional and capacity building, as well as studies on alternative sources of water, hydrogeological and hydrological assessments and mapping, water quality, and sediment loads.

Country	Project	Environmental component
Mauritius	Port Development and Environment Protection	Provides for a study to define the requirements and to design facilities for collection and disposal of land- and ship-based oily waste and to draft regulations to enforce safety in port operations and control vessels' seaworthiness; constructs facilities for the collection and pretreatment of oily waste; and supervises contracts for civil works and equipment and training to fight oil spills.
Niger	Pilot Private Irrigation Promotion	Finances the materials and part of the operating costs for the monitoring of replenishable shallow aquifers and erosion control works.
Senegal	Water Sector	Supports institutional capacity building, which will formulate a national strategy for sanitation in urban and periurban areas, and provides for the treatment and reuse of periurban municipal sewage for irrigation; prepares sewerage master plans for nineteen secondary cities and riparian areas, implements a rainwater sanitation project, and implements the Rufisque sewerage and urban drainage subproject.
Swaziland	Urban Development	Provides for the construction, rehabilitation, or expansion of existing wastewater treatment plants.

(Table continues on the following page.)

Table C.1 ***(continued)***

Country	*Project*	*Environmental component*
Tanzania	Mineral Sector Development	Helps artisans and small-scale miners to mitigate the environmental hazards of present mining methods through specific project interventions, including education and training, the promotion of simple and environmentally sensitive technologies, and improved mining and work safety measures; the monitoring of environmental aspects of mining operations and the introduction of effective regulatory measures; and support to the Regional Miners' Association to address the environmental and social problems on a self-help and self-policing basis. For medium- and large-scale industrial mining operations, develops and implements appropriate environmental policies, regulatory standards, and monitoring mechanisms. This project also provides necessary manpower, and logistical and institutional capacity to enable the government of Tanzania to implement a new environmental framework.
Asia		
Bangladesh	Gas Infrastructure Development	Supports the development of the energy sector's institutional capabilities, particularly in the areas of safety and environmental protection.
China	Shenyang Industrial Reform	Finances the implementation of hazardous waste collection and treatment facilities, an automatic wastewater monitoring network to improve enforcement of discharge standards, and technical assistance for the environmental protection bureaus to enhance their capabilities to monitor and enforce pollution control standards.

Country	*Project*	*Environmental component*
China *(continued)*	Seventh Railway	Supports the conversion from diesel to electric railcars, strengthening capabilities of the Ministry of Railways environmental staff, developing a solid waste management strategy for the Shenyang station, assessing techniques for cleaning up contaminated soil at the Taopu freight yard, treating oily wastewater at a locomotive depot, training railway environmental staff in environmental impact assessment techniques, developing a fuel additive for diesel locomotives to boost combustion and reduce emissions, preparing an environmentally sound strategy to dispose of more than 1 million Styrofoam lunch boxes per day.
	Yangtze Basin Water Resources	Improves environmental monitoring in the water sector and introduces comprehensive river basin management and nonwater subprojects on forestry, grassland development, and soil improvement.
India	Second Madras Water Supply	Supports improvements of environmental monitoring in the water sector and introduces comprehensive river basin management and nonwater subprojects in forestry, grassland development, and soil improvement.
	Tamil Nadu Water Resources Consolidation	Includes capacity building in water planning, environmental management, and research to introduce multisectoral water planning, to incorporate environmental management in all aspects of water planning, investment, and management, and to enhance research in the water sector.

(Table continues on the following page.)

Table C.1 ***(continued)***

Country	*Project*	*Environmental component*
Indonesia	Kalimantan Urban Development	Provides for investment in urban infrastructure in five cities (Banjarmasin, Balikpapan, Palangkaraya, Pontianak, and Samarinda), including drainage, human waste management, and solid waste management infrastructure. Also provides for assistance in program management and monitoring, municipal management, improved environmental and resettlement practices, institutional capacity-building for relevant agencies.
Korea, Rep. of	Pusan Urban Transport Management	Provides for transport system management actions for a bus priority corridor, as well as a congestion management component.
Pakistan	Second Private Sector Energy Development	Provides consulting services to develop environmental and resettlement assessment guidelines, as well as strengthen capabilities for monitoring compliance with these guidelines.
Thailand	Lam Takhong Pump Storage	Provides funding for computer hardware, software, and training for strengthening the environmental capabilities of the Electricity Generating Authority of Thailand.
Eastern Europe and Central Asia		
Albania	Irrigation Rehabilitation	Provides technical assistance to update the Waste Master Plan and ensure against water loss.
Armenia	Irrigation Rehabilitation	Contributes to improvements in water management by establishing pilot projects with water users' associations and assistance to update the Water Master Plan.

Country	*Project*	*Environmental component*
Azerbaijan	Greater Baku Water Supply Rehabilitation	Assists with a water demand management plan to help conserve water at the household level through a program of consumer awareness initiatives, pipe maintenance, and rationalizing of water use through meter installations and implementation of a progressive water tariff.
Estonia	Financial Institutions Development	Provides technical assistance to train financial advisers of participating commercial banks in environmental screening and assessments.
Georgia	Municipal Infrastructure Rehabilitation	Provides for infrastructure rehabilitation investment for wastewater and solid waste collection and disposal; institutional development aimed at initiating a process to develop local capabilities in environmental assessment and management; and development of strategies and action programs to control or alleviate pollution problems along the Black Sea coast.
Hungary	Budapest Urban Transport	Provides for the replacement of fifty worn-out buses with low-polluting and energy-efficient substitutes and the support of an automatic vehicle monitoring system.
Latvia	Jelgava District Heating Rehabilitation	Aims to improve environmental conditions by improving the efficiency of fuel use, facilitating the elimination of low-stack coal-fired boilers in residential areas, and reducing water wastage.
Russian Federation	Gas Distribution Rehabilitation and Energy Efficiency	Increases the efficiency of gas use by upgrading and increasing the use of customer metering and by supporting investments in more efficient equipment; provides assistance and disseminates information to industrial enterprises to evaluate and access commercial funding for energy efficiency measures.

(Table continues on the following page.)

Table C.1 *(continued)*

Country	*Project*	*Environmental component*
Turkey	Antalya Water Supply and Sanitation	Finances a new sewerage network and preliminary wastewater treatment plant.
Uzbekistan	Cotton Sub-Sector Improvement	Supports integrated pest management through equipment and technical assistance. Also supports the development of improved pesticide spraying equipment to allow the integration of biological and chemical control of pests with proper cropping practices. Provides for equipment and training to demonstrate effective ways to reduce water use in cotton production while improving land productivity and avoiding water-based damage to the environment.
Latin America and the Caribbean		
Belize	Second Power Development	Carries out environmental audits and offers cleanup loans designed for seven thermal power plants.
Bolivia	National Land Administration	Includes components that clarify and systemize the land tenure situation in Colombia by identifying overlapping land titles, alleviate land conflicts, continuing the land-titling process for small-holders, organize land tenure information, and carry out territorial planning. These activities would consistently improve the management of land resources in Bolivia.
Brazil	Recife Metropolitan Transport Decentralization	Supports the development of an integrated urban transport system as well as the design of an inspection and maintenance program for vehicle emissions; the preparation of an integrated transport policy, land use, and air quality management strategy to meet both transport and air quality targets; and the strengthening of air quality planning and monitoring of vehicle-based emissions.

Country	*Project*	*Environmental component*
Brazil *(continued)*	Ceara Urban Development and Water Resource Management	Endorses a state water resources plan, the approval of a state water law, and the creation of an integrated water resource management system.
	Belo Horizonte Metropolitan Transport Decentralization	Supports the development of an integrated urban transport system as well as the design of an inspection and maintenance program for vehicle emissions; the preparation of an integrated transport policy, land use, and air quality management strategy to meet both transport and air quality targets; and the strengthening of air quality planning and monitoring of vehicle-based emissions.
Chile	Third Road Sector	Strengthens the capacity of the Ministerio de Obras Públicas (Ministry of Public Works) for environmental impact assessment and monitoring in the road sector.
Colombia	Energy Sector Technical Assistance	Provides assistance in executing sectoral environmental assessments, formulating energy sector environmental regulations and guidelines, and promoting user awareness and participation in the definition of energy sector projects. Also provides for the development of a demand-side management strategy for efficient energy use.
Mexico	Rainfed Areas Development	Provides for rural investment in soil conservation works, the transfer of agricultural technology for soil and water conservation practices and equipment, and institutional development and training to strengthen the expertise in areas of soil conservation and the monitoring and evaluation of environmental impacts.

(Table continues on the following page.)

Table C.1 *(continued)*

Country	*Project*	*Environmental component*
Paraguay	Asunción Sewerage	Provides for the design of a control and monitoring program for sewage disposal into the Paraguay River as well as the design of a monitoring system to control hazardous and toxic chemicals from future industrial wastes.
Peru	Lima Water Rehabilitation and Management	Aims to improve the efficiency of water and sanitation delivery in the Lima-Callao metropolitan area. Initiates conjunctive use of surface and groundwater in three districts where groundwater is depleted and the risk of saline contamination is serious. Rehabilitates damaged water supply and sewerage systems and supports reforms in the legal and institutional framework of the water and sanitation sector.
Trinidad and Tobago	Water Sector Institutional Strengthening	Strengthens existing and new institutions, develops an integrated approach to water resources management, and strengthens and complements environmental planning and management. Provides funding for hydrological and hydrometeorological telemetering equipment and geographic information systems needed to improve data collection and analysis for improved water resource management.
Venezuela	Agricultural Extension	Provides for training in environmental impact of traditional and emerging agricultural production techniques and integrated pest management and other practices designed to permit more effective incorporation of ecological factors into overall farm management. Also provides for the inclusion of a specialist in environmental issues in agriculture among staff of the Foundation for Training and Applied Research for Agrarian Reform.

Country	*Project*	*Environmental component*
Middle East and North Africa		
Jordan	Agriculture Sector Adjustment	Supports efficient natural resource use by supporting reforms to restructure water sector institutions to maximize efficiency and unify planning and management, promote efficient use of existing water resources through demand management, establish control over overexploited groundwater resources, and prioritize public investments in water.
	Agriculture Sector Technical Support	Supports initiatives for measuring and managing water in the Jordan Valley, monitoring and controlling groundwater in the basin; and strengthening research and monitoring and evaluating the impacts of the country's agricultural policy adjustments.
Tunisia	Water Supply and Sewerage	Includes the construction of sewage treatment facilities in three towns and supports institutional development with a view to enhancing water resources management.

Annex D. Global Environment Facility (GEF) and Multilateral Fund for the Implementation of the Montreal Protocol (MFMP) Investments Approved in Fiscal 1995

This annex provides brief descriptions of all GEF Pilot Phase, GEF1, and MFMP investment projects managed by the Bank Group that were approved between July 1, 1994, and June 30, 1995 (tables D.1 and D.2). All Bank-approved GEF Pilot Phase, GEF1, and MFMP projects are listed in tables D.3 and D.4. In the case of GEF projects, dollar figures represent grant financing from the Global Environment Trust Fund, the core fund of the GEF. In most cases, figures represent only a portion of the total project cost. In the case of projects funded by the MFMP, figures represent grant financing from the Multilateral Fund and will pay full eligible incremental costs of phaseout activities for ozone-depleting substances (ODS). In this case, the amounts that the client and the Bank have agreed to under an umbrella grant agreement and the overall amount committed by the MFMP Executive Committee are both shown. Annex D corresponds to annex G in the full-length report.

Table D.1. GEF Projects Approved in Fiscal 1995

Country and grant amount	*Project*
Cameroon $6m grant	Biodiversity Conservation and Management: Participatory biodiversity conservation plans will be developed and implemented for eight sites in the species-rich lowland forests of southeast Cameroon and parts of the northern savanna ecosystem. The capacity of natural resource management institutions will be strengthened through financing zoological and biological inventories, by providing equipment, training, and technical assistance, and through "twinning" between the National Herbarium and the Kew Botanical Gardens and the National Herbarium of France.

Country and grant amount	*Project*
China $17.9m grant	Nature Reserves Management: In accordance with China's NEAP and Biodiversity Action Plan priorities, this project will prepare and implement management plans in five priority protected areas, train staff, fund physical investments, and work with communities adjacent to and within protected area boundaries to create incentives for sustainable use of resources. A second component will restructure a major timber industry in Changqing to promote sustainable forestry and create a core protected area of giant panda habitat, surrounded by a limited-use production-buffer zone. The project will build technical and managerial capacity of the Division of Nature Reserves through developing a national training team for biodiversity. It will fund a national nature reserve plan, equipment, and policy studies and research and will set up an information management system.
Czech Republic $2.3m grant	Phaseout of Ozone-Depleting Substances: The project will eliminate production of chlorofluorocarbons (CFCs) in the Czech Republic. It will establish a national refrigerant recovery-reclamation-recycling program. It will phase out CFCs in certain commercial, industrial, and transport refrigeration systems and introduce low- and nonozone-depleting foam technologies.
Malawi $5.0m grant	SADC Lake Malawi/Nyasa Biodiversity Conservation: Lake Malawi, Africa's third-largest lake, is a unique freshwater ecosystem, home to over 500 endemic species of fish. The project will conduct faunal surveys, identify biodiversity hot spots, prepare a conservation and management plan for the lake, recommend revisions to national environmental legislation, and fund environmental training and education activities. A parallel Canadian project (C$4.2 million) will finance research capacity-building through twinning with a Canadian institution, limnology and water quality monitoring, laboratory equipment, and public education.
Mali $2.5m grant	Household Energy: The project addresses both the demand for and supply of household energy in a country where people depend heavily on environmentally damaging woodfuels. Consumption of these fuels will be reduced by promoting the use of more efficient stoves and by substituting woodfuels with kerosene and bottled gas through technical assistance and credit to manufacturers and marketers. In parallel, forest planning and management will be strengthened through training and technical assistance.

(Table continues on the following page.)

Table D.1 ***(continued)***

Country and grant amount	*Project*
Morocco $6m grant	Repowering of Power Plant: Repowering an existing plant with a second-generation advanced combustion turbine, together with improving the quality of the materials and enhancing the scheme's performance in the power generating system of Morocco. Close monitoring of the project's impact and careful analysis of the findings will help to determine its wider applicability for use throughout the country.
OECS countries $12.5m grant	Ship-Generated Waste Management: Project will assist OECS governments in reducing pollution of international and territorial waters caused by the discharge of ship-generated solid wastes by supporting appropriate actions aimed at improving collection, treatment, and disposal of these wastes. Project includes national components consisting of the establishment of port waste-reception facilities and incremental expansion of landfill sites to handle ship-generated wastes, together with a regional component comprised of support activities and technical assistance for project management, training and education, establishment of common legal framework for ship waste management, recycling possibilities, and public awareness programs. Project activities will also protect critical habitat for the endangered Grenada dove.
Peru $5m grant	National Trust Fund for Protected Areas: The GEF will provide seed money for the Trust Fund for Conservation of Peru's Protected Areas. Annual revenue from the trust will finance management activities—including training, management plans, operational costs, salaries, awareness programs, and buffer zone alternative livelihood activities—for three key protected areas: Manu National Park and Biosphere Reserve, Noroeste Biosphere Reserve, and Rio Abiseo National Park. As the fund grows through outside contributions, additional protected areas will receive support. The Deutsche Gesellschaft für Technische Zusammenarbeit GmbH (GTZ) is providing parallel financing to support the National Institute of Natural Resources and recurrent expenses of the fund's administrative agency during 1995. The GTZ parallel grant also finances development of a master plan for all Peruvian protected areas and a two-year NGO small grants program for sustainable, integrated conservation and development projects.

Country and grant amount	*Project*
Poland $5m grant	Efficient Lighting: Through the IFC, direct subsidies will be given to manufacturers of compact fluorescent lighting. Manufacturers and wholesalers will be required to pass on full savings on to retailers, who will apply a standard percentage-based mark-up, passing savings on to consumers. This will spark demand for compact fluorescent lighting in Poland and demonstrate the financial and commercial benefits of energy-efficient lighting. The results will be large power savings and reduced emissions from coal-fired generation plants.
$25m grant	Coal-to-Gas Conversion: The project expects to demonstrate interfuel substitution and technological innovation as a means of reducing carbon dioxide emissions. GEF funds will be used to extend coal-to-gas conversions to medium-size boilers whose owners could not achieve acceptable financial rates of return without concessional financing.
Romania $4.5m grant	Danube Delta Biodiversity: The project aims to protect the Romanian delta ecosystem and contribute to the conservation of biodiversity within the delta. It will strengthen institutional capacity to monitor and manage protected areas effectively, work with local community groups to ensure sustainable use of resources, and restore some wetlands to their natural condition by testing various approaches and monitoring their impact.
Tunisia $4m grant	Solar Water Heating: The project will promote the commercialization of solar water-heating technology in the residential and tertiary sectors by conditioning the market for sustained penetration of the technology as a least-cost alternative under competitive market conditions.
Uganda $4m grant	Conservation of the Bwindi Impenetrable National Park and the Mgahinga Gorilla National Park: The project will establish a trust fund, the income from which will provide a sustainable source of funds for the management of the Bwindi Impenetrable National Park and Mgahinga Gorilla National Parks and the conservation of their biodiversity. A Trust Management Board, which represents local communities, NGOs, and the government will allocate the fund's net income to selected park management, research, and community development projects.

Table D.2. MFMP Investments Approved in Fiscal 1995

Country and grant and commitment amounts	*Project*
China $90m Grant Agreement $29m commitment	ODS Project III: The project will support from thirty to forty subprojects (about ten in each tranche) in all five ODS-user industries aimed at adoption of Grant Agreement on non-ODS technology through technical and financial assistance to enterprises for technology transfer, design, training, and implementation; assist in closures of CFCs production and halon plants; and provide technical assistance to the National Environment Protection Agency and China Investment Bank in the areas of procurement, disbursement, ODS technology,and project development and implementation. The projected average subproject size is about $1.5 million.
Egypt $2m Grant Agreement $2m commitment	IFC-MCMC Compressor: A grant of $2.1 million has been approved by the MFMP to assist the MISR compressor manufacturing company (MCMC) in converting its compressor production facility from CFC-12 compressors to HFC-134a. Originally a grant was approved for $2.8 million; however, Zanussi (Italy) will purchase 25 percent of the company as part of a financial and technical restructuring. This foreign ownership reduced the eligible funding to $2.1 million. With the financial restructuring of MCMC in Egypt, it is likely to be approved by IFC's Board—the grant agreement was signed in May 1995—and is expected to be disbursed quickly because a substantial part of the capital investments required for the conversion have already been made. After conversion is complete, it is expected that about 385 tons of CFCs will be eliminated per year.
India $50m Grant Agreement $8m commitment	ODS Project II: Two-part project: (a) supports about forty subprojects in all ODS-user and -production industries including refrigeration, air conditioning, foam, aerosols, and solvents and halons, and (b) supports non-ODS production and ODS closure projects. Each subproject will support participating enterprises in technology transfer, design, training, and implementation. Subprojects are being selected on the basis of cost-effective ODS-reduction impact. In addition, UNDP is preparing foam projects that complement the proposed Bank project.

Country and grant and commitment amounts	*Project*
India $1.25m Grant Agreement $1.25m commitment	ODS Project I: The project has two components: Shriram Refrigeration Industries, Ltd., to provide assistance for conversion of compressor manufacture from CFC-12 to HFC-134a designs and Blue Star, Ltd., to support funding for substitution of CFC-11 refrigerant with HCFC-123 in centrifugal chillers. Both components are designed to assist the participating enterprises in switching from use of ODS to non-ODS materials and technology. The subprojects represent measures that must be taken early to enable India to develop a cost-effective program to phase out ODS consumption, and they require no policy actions to be taken by the Montreal Protocol Executive Committee. Each subproject includes technology (from established technology supply relationships), design, training, equipment, and implementation of product and engineering development activities to enable each enterprise to convert to non-ODS technology.
Indonesia $17m Grant Agreement $9m commitment	ODS Project I: The project is composed of three elements: subproject preparation, subproject investments, and technical assistance. Under an Ozone Trust Fund–Project Preparation Advance, a consulting firm has been selected to prepare twenty subproject concepts and to finalize twelve for preappraisal. Additional project preparation advance funds will be requested from the Montreal Protocol Executive Committee as required. The project will support from twnty-five to thirty subprojects in several tranches and will cover all major ODS-use sectors in Indonesia, including refrigeration, air-conditioning, foams, fire protection (halons), aerosols, and solvents. Subprojects will assist beneficiaries with technology transfer, design, safety, training, and implementation. The project also includes $50,000 of technical assistance for the strengthening of UPPINDO Bank's technical capabilities in project review and supervision.

(Table continues on the following page.)

Table D.2 ***(continued)***

Country and grant and commitment amounts	*Project*
Philippines $30m Grant Agreement $12m commitment	ODS Project I: The project includes an investment component and a technical assistance (TA) component. The TA component includes an Institutional Strengthening Project to strengthen the Department of Environment and Natural Resources' capability to implement the country's ODS Phaseout Program and Technical Assistance to the Financial Agent (the Land Bank of the Philippines) to strengthen its capability to appraise and supervise subprojects in ODS refrigerant recycling and conversion to non-ODS technology in solvents, refrigeration, foam blowing, and tobacco puffing. Of the sixteen identified subprojects, ten subprojects, representing about 85 percent of the total project investment cost and about 75 percent of the total grant amount, have been preappraised. The project is expected to be complete by December 31, 1997.
Thailand $140m Grant Agreement $13m commitment	ODS Project I: The project will support a total of forty-five to fifty-five subprojects in solvents, refrigeration, air conditioning, foams, and recycling. Each subproject includes technical assistance to support participating enterprises with technology transfer, design, training, and implementation. About twenty subprojects have been identified thus far, out of which twelve subprojects have been preappraised, representing about 30 percent of the total project investment cost and about 30 percent of total project amount. Another twenty-five to thirty-five proposals will be developed during the next two years.

Country and grant and commitment amounts	*Project*
Uruguay $5m Grant Agreement $1.2m commitment	ODS Project I Investment: The project has two components. First, the investment component will provide subgrants to private firms to assist them in switching to nonozone-depleting materials and technology. This component consists of approximately ten subprojects, four of which have been already appraised by the Bank and approved by the Montreal Protocol Executive Committee. The subproject will address production process conversions, mechanical equipment and installations, recovery and recycling equipment, and retrofitting of existing facilities. Second, a TA component will strengthen the Comisión Tecnica Gubernamental de Ozono as the executing unit in supporting the Ministerio de Vivienda, Ordenamiento Territorial y Medio Ambiente to provide assistance to the beneficiaries in the identification, preparation, evaluation, and administration of subprojects. This TA component will provide funds for specialized training, office equipment, and consulting services. The project's expected completion and closing dates are December 31, 1997, and June 30, 1998, respectively.
Venezuela $3m Grant Agreement $3m commitment	ODS Project III—FAACA: The proposed project consists of retooling a heat exchanger fabrication and assembly line and related training and technical assistance costs. The assembly line conversion will permit production of mobile air conditioning systems that permit use of HFC-134a rather than CFC-12 as the refrigeration agent. A Calorimeter test laboratory for which funding was originally requested was not found to be an incremental cost and has been deleted from the project scope. Certain categories of recurrent annual investments related to model changes were also considered to be nonincremental and, likewise, are not included in the project scope.

(Table continues on the following page.)

Table D.2 ***(continued)***

Country and grant and commitment amounts	*Project*
Venezuela *(continued)*	
$.25m Grant Agreement $.25m commitment	ODS Project II Conversion of CFC-12 to HCFC-134a: The project involves the retrofitting of twenty-five existing chiller units with ten or more years of expected useful life and the installation of four new HFC-134a chillers to replace CFC-12 chillers whose useful life has expired. The retrofits are equipment-specific and have been engineered and specified by the Snyder General McQuay Co. (SGMQ), current owner of the former Westinghouse compressor technology. The four new SGMQ chillers will also replace Westinghouse chillers on a sole-supplier basis. The refitting is mechanical, involving changing of gears, seals, and gaskets.

Table D.3. GEF Pilot Phase and GEF 1 Investments Approved through Fiscal 1995

Country	*Project name (fiscal year of Bank approval)*	*Grant amount (millions of dollars)*
GEF 1 investment		
China	Nature Reserves Management (1995)	17.9
Total GEF 1 investments		17.9
GEF Pilot Phase investments		
Algeria	El Kala National Park and Wetlands Management (1994)	9.2
Algeria, Morocco, Tunisia	Oil Pollution Management Project for the Southwest Mediterranean Sea (1994)	18.3
Belarus	Biodiversity Protection (1992)	1.0
Bhutan	Trust Fund for Environmental Conservation (1992)	10.0
Bolivia	Biodiversity Conservation (1993)	4.5
Cameroon	Biodiversity Conservation and Management (1995)	6.0
China	Ship Waste Disposal (1992)	30.0
	Sichuan Gas Transmission and Distribution Rehabilitation (1994)	10.0
Congo	Wildlands Protection and Management (1993)	10.0
Costa Rica	Tejona Wind Power (1994)	3.3
Czech Republic	Phaseout of Ozone-Depleting Substances (1995)	2.3
	Biodiversity Protection (1994)	2.0
Ecuador	Biodiversity Protection (1994)	7.2
Egypt	Red Sea Coastal and Marine Resource Management (1993)	4.8
Ghana	Coastal Wetlands Management (1993)	7.2
India	Alternate Energy (1993)	26.0
Indonesia	Biodiversity Collections (1994)	7.2
Iran, Islamic Rep. of	Teheran Transport Emissions Reduction (1994)	2.0
Jamaica	Demand Side Management Demonstration (1994)	3.8
Lao PDR	Wildlife and Protected Areas Conservation (1994)	5.0
Malawi	SADC Lake Malawi/Nyasa Biodiversity Conservation (1995)	5.0
Mali	Household Energy (1995)	2.5
Mauritius	Sugar Bio-Energy Technology (1992)	3.3
Mexico	Protected Areas Program (1992)	25.0
Mexico	High Efficiency Lighting Pilot (1994)	10.0
Morocco	Repowering of Power Plant (1995)	6.0

(Table continues on the following page.)

Table D.3 ***(continued)***

Country	*Project name (fiscal year of Bank approval)*	*Grant amount (millions of dollars)*
OECD countries	Ship-Generated Waste Management (1995)	12.5
Peru	Trust Fund for Parks and Protected Areas (1995)	5.0
Philippines	Conservation of Priority Protected Areas (1994)	20.0
	Leyte-Luzon Geothermal (1994)	30.0
Poland	Forest Biodiversity Protection (1992)	4.5
	Efficient Lighting (1995)	5.0
	Coal-to-Gas Conversion (1995)	25.0
Romania	Danube Delta Biodiversity (1995)	4.5
Seychelles	Biodiversity Conservation and Marine Pollution Abatement (1993)	1.8
Slovak Republic	Biodiversity Protection (1994)	2.3
Thailand	Promotion of Electricity Energy Efficiency (1993)	9.5
Tunisia	Solar Water Heating (1995)	4.0
Turkey	In-Situ Conservation of Genetic Biodiversity (1993)	5.1
Uganda	Conservation of the Bwindi Impenetrable National Park and the Mgahinga Gorilla National Park (1995)	4.0
Ukraine	Transcarpathian Biodiversity Protection (1994)	0.5
	Danube Delta Biodiversity (1994)	1.5
Wider Caribbean	Wider Caribbean Initiative for Ship-Generated Waste (1994)	5.5
Total GEF 1 investments		17.9
Total GEF Pilot Phase investments		362.3
Total Fiscal 1992–95		380.2

Table D.4. Montreal Protocol Investment Operations Approved by the World Bank, Fiscal 1992–95
(millions of dollars)

Fiscal year	*Country*	*Project*	*Umbrella*	*Committed*
1992	Mexico	MAC Recycling and Aerosols		0.18
1992	Malaysia	Halons and MAC Recycling		1.63
1992	Philippines	ODS Phaseout Engineering		0.39
1992	Thailand	ODS Phaseout Engineering		0.39
Total				2.59
1993	Chile	ODS Phaseout		1.21
1993	Mexico	Ozone Protection Policy		4.00
1993	Venezuela	ODS Phaseout I–Plasticos Molanca		1.30
Total				6.51
1994	Brazil	ODS Phaseout I	10.90	5.14
1994	China	ODS Phaseout I		6.93
1994		ODS Phaseout II		4.47
1994	Ecuador	ODS Phaseout I		1.57
1994	Jordan	TA and Investment Project		1.50
1994	Turkey	ODS Phaseout I		6.17
1994	Tunisia	ODS I		1.79
Total			10.90	27.57
1995	China	ODS III	90.00	29.45
1995	Egypt	MCMC Compressor (IFC)		2.10
1995	India	ODS I		1.25
1995		ODS II	50.00	8.12
1995	Indonesia	ODS I	17.00	8.68
1995	Philippines	ODS I	30.00	11.73
1995	Thailand	ODS Phaseout I	40.00	12.50
1995	Uruguay	ODS I	5.00	1.23
1995	Venezuela	Chiller Retrofits		0.25
1995		FAACA		3.10
Total			232.00	78.41
Total			242.90	115.08

Bibliography

Titles published by the World Bank can be obtained from the bookstores at the World Bank offices in Washington, D.C., and Paris or through the World Bank's authorized commercial distributors and depository libraries throughout the world. Abbreviations used are as follows: EDI, Economic Development Institute (of the World Bank); GEF, Global Environment Facility; and World Bank Environment Department divisions (ENVGC, Global Environment Division; ENVLW, Land, Water and Natural Habitats Division; ENVSP, Social Policy and Resettlement Division; ENVPE, Pollution and Environmental Economics Division). Titles preceded by an asterisk are available by writing directly to the department named. The bibliographical entries are as they appear in the full-length report.

World Bank. 1991a. *The Forest Sector.* A World Bank Policy Paper. Washington, D.C.

———. 1991b. *Urban Policy and Economic Development: An Agenda for the 1990s.* A World Bank Policy Paper. Washington, D.C.

———. 1993a. *Energy Efficiency and Conservation in the Developing World.* A World Bank Policy Paper. Washington, D.C.

———. 1993c. *Water Resources Management.* A World Bank Policy Paper. Washington, D.C.

———. 1993e. *The World Bank's Role in the Electric Power Sector: Policies for Effective Institutional, Regulatory, and Financial Reform.* A World Bank Policy Paper. Washington, D.C.

*———. 1994i. "Review of Implementation of the Forest Sector Policy." ENVLW and Agriculture and Natural Resources Department, Washington, D.C.

*———. 1995g. "Industrial Pollution, Prevention and Abatement Handbook." Draft. Environment Department, Washington, D.C.

———.1995i. *Monitoring Environmental Progress: A Report on Work in Progress.* Washington, D.C.

*———. 1995j. " National Environmental Strategies: Learning from Experience." ENVLW, Washington, D.C.

*———. Forthcoming b. "Second Review of Environmental Assessment." ENVLW, Washington, D.C.

*World Bank, TWU (Transport, Water, and Urban Development Department). 1995. "Sustainable Transport: A Sector Policy Review." Draft. Washington, D.C.

World Bank, World Conservation Union, and Great Barrier Reef Marine Park Authority. 1995. *A Global Representative System of Marine Protected Areas,* 4 vols. Washington, D.C.